34.00

THE WAR ON WASTE

BY VALERIE BODDEN

CREATIVE C EDUCATION

Published by Creative Education
P.O. Box 227, Mankato, Minnesota 56002
Creative Education is an imprint of
The Creative Company
www.thecreativecompany.us

Design and production by The Design Lab
Art direction by Rita Marshall
Printed by Corporate Graphics in the
United States of America

Photographs by Alamy (Peter Arnold, Inc.,
Phillip Augustavo, Borderlands, Ray Roberts,
Jim West), Corbis (Peter Essick/Aurora
Photos, Gilles Sabrié, Scheufler Collection),
Dreamstime (Irina Drazowa-Fischer), iStock-
photo (Mark Connors, Sebastien Cote, Antonio
D'Albore, Jaap Hart, Jeffrey Heyden-Kaye,
Stefan Klein, Peter Nad, Johannes Norpoth,
Peeter Viisimaa, Roger Whiteway)

Library of Congress
Cataloging-in-Publication Data
Bodden, Valerie.
The war on waste / by Valerie Bodden.
p. cm. — (Earth issues)
Includes bibliographical references and index.
Summary: An examination of the various
types of waste produced by people, exploring
how much of it exists and where it goes, as
well as how waste can be remade into useful
items to contribute to a healthier planet.
ISBN 978-1-58341-985-4
1. Refuse and refuse disposal--Juvenile litera-
ture. I. Title. II. Series.

TD792.B63 2010
363.72'8—dc22   2009028053

CPSIA: 120109 PO1091
First Edition
9 8 7 6 5 4 3 2 1

Table of Contents

Everything human beings need to survive—air to breathe, food to eat, water to drink—is found on Earth, and on Earth alone. Yet the very planet that sustains human life has come under threat because of human activities. Rivers are drying up as people divert water for their own use. Temperatures are warming as greenhouse gases such as carbon dioxide trap heat in the **atmosphere**. Species of plants and animals are disappearing as people destroy essential habitats. And the rate of many such changes appears to be accelerating. "If I had to use one word to describe the environmental state of the planet right now, I think I would say precarious," said population expert Robert Engelman. "It isn't doomed. It isn't certainly headed toward disaster. But it's in a very precarious situation right now."

With Earth's population approaching 6.8 billion people, one of the environmental issues that affects nearly every corner of the globe is waste disposal. In fact, at the first-ever Earth Summit held in 1992, 60 percent of nations reported that disposing of solid waste was one of their biggest environmental concerns. From cooking to cleaning, working to playing, nearly every activity in which human beings engage produces waste. Although some of that waste is natural (leftover food, for example), other items, such as chemicals or heavy metals, are hazardous to human health. But how much waste do we produce, and where does it all go? Does throwing old, broken, or unwanted items in the trash really matter? And what can we do to produce less trash?

Waste, of course, is nothing new. People have been producing waste since the beginning of civilization. At first, **nomadic** tribes simply discarded their waste, which was **organic**, to be decomposed by nature. Later, people began to dig holes in the ground to dispose of their garbage. In fact, archaeologists have found evidence that homes built around 4,500 years ago in the city of Mohenjo-daro (in present-day Pakistan) were built with garbage chutes and trash containers. Around 400 B.C., the city of Athens, Greece, made a law requiring garbage to be dumped at least a mile (1.6 km) from the city, and later, the Roman Empire established a garbage collection service. By the Middle Ages (A.D. 476–1453), however, garbage collection had lost its popularity in Europe, and people began to throw their waste into streets and rivers, leading to unsanitary conditions and the spread of disease. In 17th- and 18th-century colonial America, too, garbage was thrown into the streets, where pigs and dogs dug through the waste in search of a meal.

CHAPTER ONE

A World of Waste

The amount of waste that was tossed away then was small, though, compared to the amount of waste that exists today. That's because food scraps were fed to farm animals, broken items were fixed, old clothes were used as rags or quilts, and other items were traded to local **peddlers**. By the end of the 18th century, however, the **Industrial Revolution** had introduced new, mass-produced goods, and with city populations continuing to grow, trash piles became eyesores—as well as sources of disease. As a result, cities in both Great Britain and America gradually began to pass laws governing trash disposal and to provide refuse collection systems. Some cities dumped their collected waste into oceans, lakes, or rivers; others burned their garbage or placed it in large open dumps.

Garbage collection services have not evolved far from the days of carts picking up curbside waste in the early 20th century.

Hazardous waste in the form of used cans of paint presents a challenge to homeowners when it comes to proper disposal.

Ironically, the introduction of trash collection services led people to throw more items away. By the beginning of the 20th century, the availability of new, cheaper products also meant that people were less likely to repair a broken item when they could buy a new one. Soon, people were throwing away perfectly good objects because they simply weren't new anymore or had gone out of style. By the 1950s, people in industrialized countries—especially the United States—were characterized by an attitude of consumerism, or buying new items simply for the sake of buying. Many of these items were designed for short-term or one-time use, earning America the nickname "the throwaway society." Today, consumerism has spread around the world, as people in **developing countries** such as China and India have also begun to desire the consumer goods popular in wealthier countries.

Today's vast quantities of waste come from several sources, including households, agriculture, manufacturing, and other industries. In fact, the largest producers of waste are agriculture, construction, and mining. Agricultural waste can often be plowed back into the ground, and mining waste, known as tailings, is often disposed of close to mines (despite the fact that it is sometimes toxic). Other industries also generate large amounts of waste in their production processes. The non-hazardous portion of this waste is known as industrial non-hazardous waste. Because industrial non-hazardous waste is regulated by a number of different agencies, it is difficult to determine exactly how much of this waste is produced each year, but in the U.S., the Environmental Protection Agency (EPA) puts the figure around 7.6 billion tons (6.9 billion t). In addition, U.S. industry produces about 46.7 million tons (42.4 million t) of hazardous waste each year. Hazardous waste is defined as any type of waste that is **corrosive**, ignitable, **reactive**, or toxic.

Wasteful Behavior

Not all recyclable materials are created equal. Metal and glass can both be recycled endlessly without losing quality. Paper, on the other hand, can generally be recycled only three or four times before its fibers become too short to use. Much recycled paper is turned into thin boxes or paper towels. Plastic has a limited recycling life as well. Mixed plastics can be used to make plastic wood or toothbrush handles (neither of which are recyclable), and one of the most commonly recycled plastics, polyethylene terephthalate, or PET (used for soda bottles and designated with the number one), can be made into products such as sleeping-bag fill, carpeting, and fleece jackets.

In poorer communities such as this one in the African country of Ghana, materials are often reused instead of thrown away.

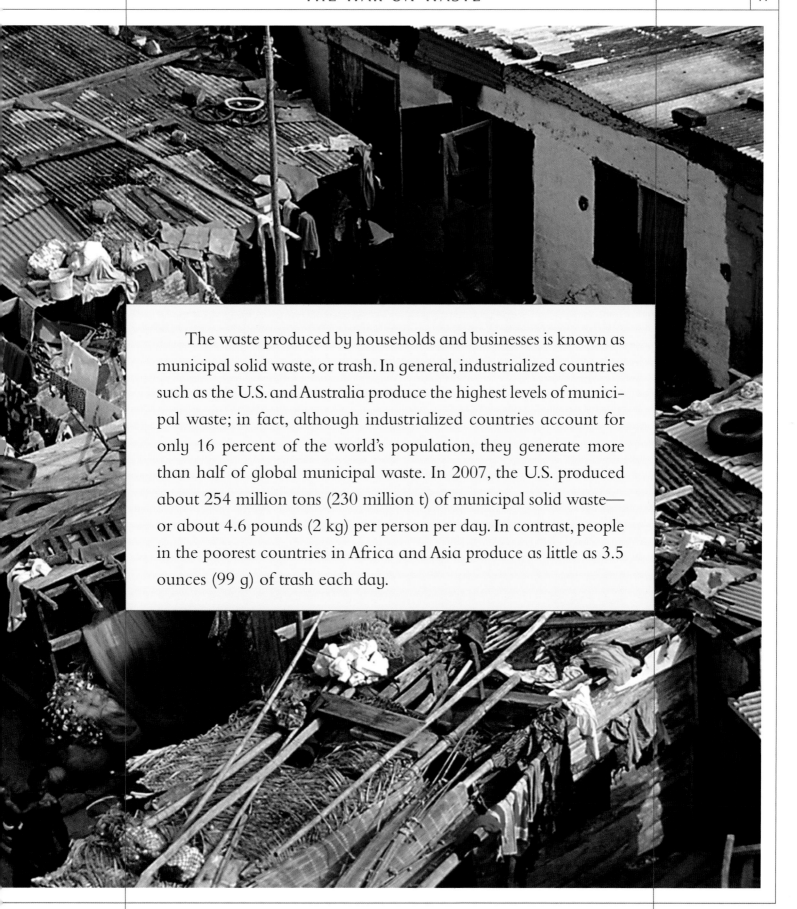

The waste produced by households and businesses is known as municipal solid waste, or trash. In general, industrialized countries such as the U.S. and Australia produce the highest levels of municipal waste; in fact, although industrialized countries account for only 16 percent of the world's population, they generate more than half of global municipal waste. In 2007, the U.S. produced about 254 million tons (230 million t) of municipal solid waste—or about 4.6 pounds (2 kg) per person per day. In contrast, people in the poorest countries in Africa and Asia produce as little as 3.5 ounces (99 g) of trash each day.

Wasteful Behavior

Many of the decisions we make while shopping can have a direct impact on how much waste ends up in landfills and incinerators. The biggest thing we can do to reduce the amount of waste we produce is to ask ourselves whether or not we really need to buy a product that will eventually end up as waste. When we do need to make a purchase, we can steer clear of disposable items such as paper plates and instead buy long-lasting reusable goods. We can also buy products made out of recycled materials and carry reusable shopping bags.

In developing nations, the majority of municipal solid waste is organic, consisting largely of food scraps. The trash in industrialized nations includes organic items such as food waste and yard trimmings, too, but it also contains a large proportion of man-made products, such as paper, chemicals, plastics, metals, and glass. Some of the trash—including food scraps, yard trimmings, and paper—is biodegradable, meaning that it can be broken down by bacteria into elements such as carbon. Other items, however, do not break down so easily. Plastics, for example, do not decompose through organic processes, although they photodegrade, or break down into smaller and smaller pieces when exposed to sunlight. No one knows how long it takes for plastics to completely break down, but some scientists estimate that it could take hundreds of years. This can be a problem, because plastics are among the fastest-growing types of waste in many parts of the world, including in some African nations, where plastic shopping bags have replaced traditional baskets. In some countries, more than half of all plastic waste consists of packaging materials, which are thrown away soon after purchasing a product.

Along with plastics, electronics are a fast-growing segment of the municipal waste stream, representing five percent of all municipal solid waste in the world. According to the United Nations Environment Programme (UNEP), 22 to 55 million tons (20–50 million t) of electronic waste, known as e-waste, are generated globally each year. And, with the average computer having a life span of only two to five years, e-waste rates are expected to continue to soar.

China is classified as a "developing nation," but it is more industrialized than other such countries and has more waste to show for it.

With so much waste generated across the globe, waste disposal has become a major environmental issue in many parts of the world. In many countries, dumping waste in landfills is seen as the cheapest and easiest way to get rid of garbage, and landfills can be found throughout the U.S., Canada, Great Britain, and Australia. In China, 65 percent of major cities are surrounded by landfills, and Beijing alone has 13 landfill sites. In the past, most landfills were simply large open holes in the ground filled with garbage, and open dumps remain a major method of garbage disposal in many Asian and African countries. In the industrialized world, however, modern landfills are lined with layers of gravel, clay, soil, sand, plastic, and **geotextiles**, and waste is covered with soil or an alternative cover such as shredded tires or crushed glass. Although landfills are likely to continue as one of the most common options for garbage disposal into the near future, finding places to build new landfills is becoming increasingly difficult. In some densely populated countries, especially those in Europe, little land is available for the development of new landfills. In other places, public opposition often prevents landfills from being built.

CHAPTER TWO

Waste Disposal

Landfills are not the only option for waste disposal. In some countries, especially where space for landfills is limited, incineration (burning waste in a high-temperature furnace) is the preferred method of garbage disposal. Japan, for example, burns its garbage in 2,800 incinerators across the country, and Norway incinerates nearly one-third of its garbage. Incineration is more expensive than landfill dumping, but it can reduce the volume of waste by 90 percent. Although the leftover ash (which sometimes

Before 1990, German incineration plants emitted harmful gases, but they have since modified their equipment to meet clean air standards.

Wasteful Behavior

Some cities and countries have set for themselves the ambitious goal of becoming "zero waste" societies, in which no waste is sent to landfills or incinerators. The state of California, the country of New Zealand, and regions of Australia and Canada have all committed to zero waste. In some places, zero waste is seen as a literal goal, while in others it is simply a guiding principle to encourage citizens to reduce the amount of waste they generate. The Australian Capital Territory, for example, had set a goal of recovering 95 percent of its waste for recycling and composting by 2010.

Antifreeze used in cars is toxic to land and water life forms and must be recycled and disposed of at an official collection center.

To safely dispose of used car batteries (which are a form of hazardous waste), many people take them to an auto supply store or a repair shop.

contains toxic elements) must still be dumped in a landfill, sometimes countries use it to make cement building blocks or to underlay roads instead. In many countries, incinerators have become a source of energy, as well as a method of waste disposal. As the waste burns, it heats a boiler, which produces steam that turns a generator, creating electricity for surrounding homes and businesses.

Hazardous waste is also often burned in specially designed incinerators or buried in certain landfills. If it has been treated first, it can be disposed of like regular waste. Most hazardous waste is created during the production of chemicals and plastics, but the dry-cleaning, furniture manufacturing, pharmaceutical, construction, and printing industries also generate small quantities of hazardous waste. In addition to the hazardous waste generated by industry, individuals produce small amounts of hazardous waste also. This is known as household hazardous waste and includes items such as paints, cleaners, **pesticides**, and antifreeze. Although many communities arrange special collections for household hazardous waste, hazardous substances often find their way into landfills and incinerators. In addition, many wastes that were not originally considered hazardous were dumped in landfills before their dangers were realized, and there they remain today.

Whether hazardous or not, not all waste is disposed of properly. Sometimes, individuals or companies are unwilling to pay the high costs of disposing of waste in appropriate landfills or incinerators and instead illegally dump it. In Los Angeles, for example, 121,000 tons (110,000 t) of solid waste is dumped illegally every year. And in Japan, a disposal company illegally dumped hazardous waste on the shores of the island of Teshima for 25 years, resulting in a cleanup operation costing more than $300 million. In the U.S., more than 1,250 locations are so contaminated with hazardous waste that they have been placed on the EPA's Superfund National Priorities List (NPL)—a list of the worst hazardous waste sites in the country. Another 40,000 or so Superfund sites not contaminated enough to make it onto the NPL have also been determined to pose a risk to human health or the environment. Many Superfund sites are located on the grounds of former landfills or metal mines (where exposure to hazardous heavy metals such as lead and cadmium can damage the nervous system, kidneys, or lungs).

Waste products from a Montana mine resulted in the designation of the Silver Bow Creek/ Butte Area Superfund site in 1983.

Wasteful Behavior

In 2006, a ship carrying waste from the Dutch company Trafigura transported more than 500 tons (454 t) of liquid toxic waste to Africa's Ivory Coast. The Ivorian firm Compaigne Tommy dumped the waste on open spaces across the city of Abidjan. In the following weeks, more than 100,000 residents flooded hospitals, complaining of breathing problems and other ailments. At least 17 people died. Experts from the United Nations determined that the waste contained the gas hydrogen sulfide, which is dangerous in concentrated amounts. In October 2008, the owner of Compaigne Tommy was sentenced to 20 years in prison for poisoning, and a shipping agent received a 5-year sentence.

Workers in China spend hours picking apart e-waste from scrap computers that are shipped there from America and other countries.

Rather than disposing of hazardous waste in their own country—whether legally or illegally—many industrialized nations have decided to export their hazardous waste to developing countries, where environmental laws tend to be lax and disposal costs usually low. One ton (0.9 t) of toxic ash that would cost $1,000 to dispose of in the U.S. can be dumped in the West African nation of Guinea for $40, for example. Hazardous waste is not the only type of waste shipped to developing nations for disposal. E-waste products such as cell phones and computers—which often contain dozens of different materials, including many dangerous heavy metals—are commonly exported from Europe, the U.S., and Japan to countries such as China, India, Pakistan, Vietnam, and Nigeria to be recycled.

Despite the fact that some developing nations import e-waste for disposal, many of these same countries lack even the most basic garbage

Windblown plastic bags make for unsightly trash when they get stuck in trees, bushes, and other parts of the natural landscape.

Wasteful Behavior

In many nations, plastic shopping bags are one of the most common forms of litter—the bags are called "white pollution" in China and the "national flower" in South Africa. Some countries have taken steps to reduce plastic bag litter, however. In 2002, Ireland instituted a tax on plastic shopping bags, and as a result, plastic bag use fell by 90 percent. China also started charging shoppers for plastic bags in 2008, despite the fact that the world's largest plastic bag manufacturer—located in China—closed as a result. Bangladesh outlawed the bags altogether in 2002 after suffering disastrous floods from bag-clogged drains.

collection and disposal services. In Accra, the capital of Ghana, for example, home collection of solid waste is available to less than 15 percent of the population; other households take their waste to communal collection points, throw it onto waste mountains in open spaces, or toss it into waterways. In Jakarta, Indonesia, 70 percent of residents' daily waste is thrown into the city's canals, which empty into the Angke River. The trash is so thick in some parts of the river that locals can actually cross the waterway by jumping from one large piece of trash to another.

While some governments have failed to offer centralized garbage collection systems to their citizens, there is one type of waste that no government in the world has yet been able to dispose of permanently: spent **nuclear** fuel. This consists of the **radioactive** remains of the fuel used by nuclear power plants, and, along with other long-lived radioactive wastes such as those created during the production of nuclear weapons, it can remain radioactive for 10,000 years or more. Even low-level radioactive waste, such as that produced by X-ray machines, can be difficult to dispose of. Currently, low-level waste is often placed into large drums and buried, while waste from the world's more than 400 nuclear power plants is kept largely in temporary storage at nuclear reactor sites.

With Jakarta's population at almost nine million and no system of waste disposal, it is no wonder that the city's waterways are full of trash.

No matter what methods are used to dispose of it, much of the world's waste has the potential to harm human health. Even waste that is buried in landfills is not completely isolated from the surrounding environment. As the waste in landfills decays and mixes with rainfall runoff, it creates leachate, or "garbage juice," a sometimes toxic brew that can contain liquefied rotting food, **pathogens**, pesticides, paint, and medical waste. Despite the fact that modern landfills are enclosed by a complex liner and also contain leachate collection pipes, which drain the leach-ate to collection ponds and wastewater treatment plants, some environmentalists argue that even the best liners will eventually leak. In France, landfills with 16-foot-thick (4.9 m) liners leaked after being damaged by chemicals found in leachate. In addition, large quantities of leachate flow from unlined landfills. Leachate threatens groundwater supplies, as it can seep through the soil and into local **aquifers**. This leachate can contain heavy metals such as cadmium, mercury, and lead.

CHAPTER THREE

Waste Worries

Groundwater can also be polluted when heavy metals or other contaminants such as **volatile organic compounds** or pesticides from hazardous waste sites seep into the ground. In fact, 85 percent of the more than 1,250 sites on the NPL in the U.S. have contaminated nearby groundwater supplies. In Africa and the Middle East, huge stockpiles of outdated pesticides have leaked into water supplies. Even litter can cause water pollution because plastic can leach phthalates—chemicals used to make plastics soft and flexible.

Waste-derived contaminants in water supplies can harm human health in many ways. Phthalates can disrupt the **endocrine system**

Although landfills are now required to treat leachate or to dig up the site entirely, such procedures are not always followed.

Wasteful Behavior

Although some decomposition occurs when garbage is first placed in a landfill, researchers who have studied the contents of landfills as part of "The Garbage Project" discovered that the process of decay slows dramatically soon afterward. The Garbage Project members found newspapers from the 1970s that were intact and readable, along with 40-year-old hot dogs that still looked edible. Landfills that are kept dry with covers and liners are especially good at preserving waste. Because of this, some waste management experts are looking into the development of bioreactors—landfills that circulate collected leachate back into the garbage heap in order to speed the process of decay.

and damage the liver. Cadmium, lead, and mercury can cause lung, kidney, nervous system, reproductive, or stomach problems. Other health issues associated with groundwater contamination from hazardous waste include cancer, low birth weight, and birth defects.

Air pollution is another common side effect of both landfills and incinerators. As waste in landfills decays, it produces huge quantities of the greenhouse gases carbon dioxide and methane. Although modern landfills in many countries are equipped with methane collection systems that burn off the methane generated by rotting garbage, some estimates hold that these systems collect only 75 percent—and possibly as little as 10 or 20 percent—of the methane produced. In addition, uncovered and unregulated landfills release methane freely, making them a significant contributor to global warming, since methane traps 20 times more heat in Earth's atmosphere than does carbon dioxide.

The methane and carbon dioxide that escape from landfills into the atmosphere carry with them volatile organic compounds such as acetone, benzene, and vinyl chloride, which are blown to homes and businesses nearby. Breathing in landfill gases has a number of health implications. A 1998 study discovered that women who lived within 250 feet (76 m) of several landfills in New York

Some waste companies can convert methane collected in landfills into electricity at gas-to-energy facilities located onsite.

Wasteful Behavior

Before it closed, the Fresh Kills Landfill on New York City's Staten Island was the world's largest landfill—so large, in fact, that it could be seen from outer space. Although Fresh Kills officially closed on March 22, 2001, after 53 years of accepting waste, the landfill temporarily reopened to receive debris from the World Trade Center attacks of September 11, 2001. Today, the city of New York is involved in a decades-long project to turn Fresh Kills into a park containing sports fields, trails for horseback riding and hiking, and locations for outdoor dining, along with wild habitats such as salt marshes, prairies, and forests.

The smoke from the chimney of an incinerator is a visual reminder of all the gases and particles such a disposal method releases.

were four times more likely to develop bladder cancer or leukemia than the general population. In addition, the volatile organic compounds found in landfill gases can damage the cardiac and reproductive systems and cause low birth weight, birth defects, and developmental disabilities in babies. Other effects of inhaling landfill gases include nausea, vomiting, headaches, and breathing difficulties. In addition, if methane emissions are not properly controlled, the gas can leak into nearby buildings and homes, where it can cause an explosion if ignited by a spark.

Burning waste in incinerators also creates many air pollutants, including the greenhouse gases nitrous oxide and sulfur dioxide, hydrochloric acid, trace heavy metals, and highly toxic compounds known as dioxins. Although modern incinerators are equipped with **scrubbers** to treat waste gases, some environmentalists argue that the scrubbers still allow minute quantities of pollutants to escape into the atmosphere—and even tiny amounts of some of these substances can be dangerous. For example, dioxins, which are created from burning certain kinds of plastics, are labeled by some scientists as the most toxic chemicals ever produced by people. They can accumulate in the human body, causing cancer, **infertility**, and birth defects, and damaging the nervous and immune systems.

While disposing of waste in landfills and incinerators can cause health problems, a lack of proper waste disposal can also endanger human health. Rotting waste can carry pathogens or attract disease-carrying animals such as insects and rodents, leading to the spread of diseases such as cholera, typhoid, and plagues. In Accra, Ghana, uncollected garbage blocks drains and creates stagnant waters that serve as breeding grounds for malaria-carrying mosquitoes; malaria has become the number-one cause of death in the city.

Despite the fact that living near open piles of garbage can be dangerous, in some parts of the world, people actually make their homes at dumpsites. These are the world's garbage pickers. Representing one percent of the world's **urban** population, garbage pickers scavenge through trash searching for items that can be exchanged for money, repaired and resold, or even eaten. Garbage pickers, many of them children, face the constant threat of infection, and diseases such as typhoid, tetanus, dysentery, and cholera are common. Toxins are also a constant danger, as are loose piles of trash. In 2005, a waste landslide at the Cimahi dump outside Jakarta, Indonesia, killed 143 people.

Rather than picking through the waste in landfills, some people in developing nations

Garbage pickers are numerous in the streets of Calcutta, India, and often live on the streets themselves or in fragile huts.

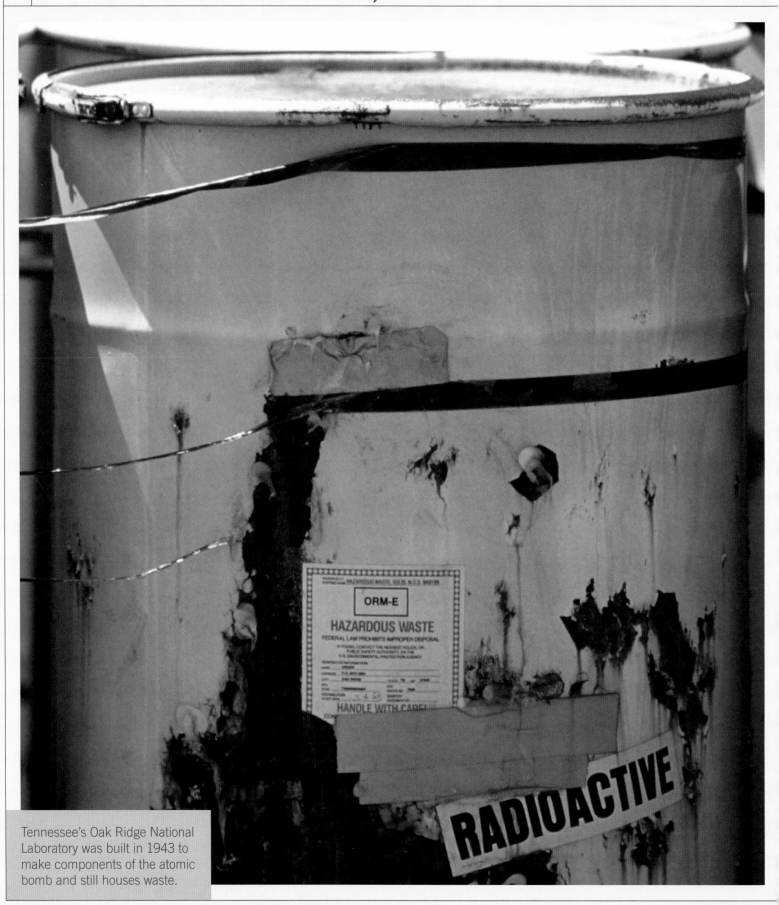

Tennessee's Oak Ridge National Laboratory was built in 1943 to make components of the atomic bomb and still houses waste.

are employed in the business of "recycling" computers from around the world. This low-tech industry involves many dangerous practices, including burning computer parts over open flames, which creates ash laced with dioxins; breaking open lead-filled computer monitors with hammers; and opening printer cartridges without protective masks or gloves. The leftover toxic waste, which includes lead, plastic, and chromium, is often dumped into nearby fields, rivers, and **irrigation** canals. Because of e-waste, the water in the city of Guiyu, China, is contaminated with lead, chromium, barium, copper, tin, and cadmium at levels hundreds or thousands of times higher than those determined to be safe by the EPA. The city's residents face high rates of birth defects, bacterial and blood diseases, and respiratory problems.

While people in the developing world face the dangers of e-waste firsthand, some people in industrialized countries worry about the hazards of radioactive waste. Although most low-level waste is buried in underground barrels, the barrels can corrode due to environmental conditions, which could allow radioactive materials to leak into water supplies. Exposure to low levels of radiation for a short time is believed to pose little danger, but exposure to high-level radiation (and possibly prolonged exposure to low-level radiation) can cause many types of cancer and birth defects and decrease the body's ability to fight off infections.

Environmentalists have offered a number of solutions to the waste problem. Most of these solutions are centered on the "three Rs" of waste disposal—reduce, reuse, and recycle. Although it is often overlooked, the most obvious way to reduce the amount of waste we produce is to buy fewer items. When consumers purchase fewer products, they not only reduce the amount of garbage in the municipal solid waste stream, but they also help to decrease the production of industrial waste, since for every pound (0.45 kg) of product made, 32 pounds (14.5 kg) of waste are generated.

CHAPTER FOUR

Reduce, Reuse, Recycle

Reusing products can also help to reduce the amount of trash that needs to be disposed of. In some cases, items can be reused many times for the same purpose. In China, for example, reusable chopsticks are helping to cut down on the 45 billion pairs of disposable ones that find their way into landfills every year. Even products that cannot be reused for the same purpose can often be put to new use. In the developing world, old soda cans are often turned into cups, and food cans become kerosene lanterns.

Many items that cannot be reused can be recycled, or processed into new products. Paper, glass, metals, and plastics are collected for recycling in many nations. South Korea has one of the highest recycling rates in the world, at nearly 50 percent; Mexico is at the other extreme, recycling less than 5 percent of its waste. In many developing countries, official recycling programs are rare, but garbage pickers play a large part in removing valuable materials from garbage dumps to be reused.

Residents of the developing world are also often involved in recycling e-waste, but due to the dangers they face, environmentalists are calling for new measures to be put into place that will

Brazilian workers sort out recyclables from other waste on conveyor belts, a time-consuming but environmentally worthwhile task.

Wasteful Behavior

Citizens in the city of Curitiba, Brazil, separate their waste into two categories—organic and inorganic. The waste in most of the city is picked up by trucks, but in the poorer sections, steep, unpaved roads prevent truck travel, so residents of these areas bring their waste to community centers. In return, they receive bus tickets or locally grown foods. Recent immigrants, recovering alcoholics, and disabled people are hired to sort through the trash for recyclable materials, which are sold to local industries. As a result of this program, Curitiba boasts a 67 percent recycling rate—one of the highest in the world.

ensure that e-waste recycling is done only in facilities equipped to handle such waste. To this end, the European Union's (EU's) Waste Electrical and Electronic Equipment (WEEE) Directive requires electronics manufacturers selling goods in the EU to take back their products for safe recycling when consumers are done with them. In many countries, such as Switzerland, the equipment is sent to high-tech electronics recycling facilities, where machines do much of the work.

The WEEE Directive is known as an extended producer responsibility (EPR) law because it requires manufacturers to take responsibility for the waste created at the end of a product's useful life. In some countries, EPR laws have also been aimed at packaging materials. In Germany, for example, the Green Dot program requires manufacturers to receive and recycle used packaging, including shipping cartons and pallets, cardboard boxes, and even toothpaste tubes.

The WEEE Man sculpture in London is 23 feet (7 m) tall and composed of the average amount of e-waste generated in a person's lifetime.

While recycling can go a long way toward removing paper, plastic, glass, metal, and electronics from the waste stream, a form of natural recycling called composting can help to remove organic items such as food scraps and yard trimmings from landfills and incinerators. When items are composted, they are allowed to decompose until they form a thick, rich **humus**, which can be used as fertilizer or mulch. In many countries, such as the Philippines, compostable materials are taken to municipal composting sites. Home composting in backyard bins is also growing in popularity in many places around the world.

Even with recycling and composting programs in place, many items still make their way to incinerators and landfills every day. Many people believe that we need to make better use of this waste by harvesting it for energy production. Already, incinerators around the world generate energy as they burn their waste in a process known as waste-to-energy. While concerns about the emissions from such plants remain, new technologies that use **plasma** to vaporize garbage at temperatures of 10,000 °F (5,538 °C) promise to produce fewer emissions. Energy can also be derived from decomposing waste in landfills, as the methane gas generated by decomposition can be collected and used to make electricity or to create **liquefied natural gas**.

From commercially produced bins to do-it-yourself contraptions, methods of home composting are becoming more widely used.

Wasteful Behavior

Land is not the only place where litter can be found. In fact, the Pacific Ocean contains the largest garbage patch in the world. Twice as big as the state of Texas, the Eastern Garbage Patch is spread out over a large region located between the coast of California and the islands of Hawaii. Ocean currents carry garbage—including sneakers, plastic bottles, basketballs, and toothbrushes—to this region. Masses of fishing nets and ropes are also tangled among the litter, which often washes up on remote beaches in the Hawaiian Islands, some of which are now piled high with trash.

Hawaii is often known for its white sandy beaches, but in reality, many of its shores are being clogged with trash from the ocean.

Of course, in some parts of the world, the problem lies not in making better use of garbage but in finding somewhere safe to dispose of it in the first place. In many countries where the local government has failed to provide adequate waste management systems, community members have stepped up to solve some of the worst waste problems. In Kampala, Uganda, for example, young people have created a community composting program, and in Ghana, schoolchildren have established community cleanup campaigns.

Sometimes cleaning up waste isn't as simple as picking up litter. In the case of hazardous waste, long-term, intensive cleanup operations are often necessary. Removing dangerous toxins can involve treating contaminated soil or water to reduce the toxicity or volume of hazardous waste they contain; erecting physical barriers such as caps, or covers, on contaminated sites to stop the spread of contamination and prevent humans from coming into contact with it; or completely removing the contaminated soil or water from an area. On the island of Teshima in Japan, for example, about one-third of the 500,000 tons (453,600 t) of hazardous waste illegally dumped along the shore was removed as of 2007; the rest was covered with liners to prevent pollutants from coming into contact with rainwater, which could cause the contaminants to leach and spread.

Nuclear waste poses its own hazards, and many governments are looking for ways to dispose of it in deep geological repositories—storage facilities isolated far underground. Although several countries, including Germany, Sweden, Finland, and Norway, have established shallow repositories for low- and medium-level radioactive waste, so far, the U.S. is the only country with a deep geological repository for long-lived nuclear waste. The Waste Isolation Pilot Plant, located near Carlsbad, New Mexico, has been receiving waste (largely in the form of contaminated clothing and equipment) from nuclear weapons production and research facilities since 1999. Progress on a site to store spent fuel from nuclear power plants has been slower, although a repository is scheduled to open at Yucca Mountain in south-central Nevada as early as 2020. Other countries are also investigating potential sites for storing spent nuclear fuel. Some people fear, however, that transporting nuclear waste to deep geological repositories threatens large segments of the population, and others argue that earthquakes could damage underground repositories. So far, however, no one has figured out an alternative way to permanently store this radioactive waste.

From nuclear and hazardous waste to electrical appliances and product packaging, the amount of waste in the world grows every day. Fortunately, there are things we can do now to prevent future generations from having to deal with our mess. By changing our perceptions of waste, we can learn to see used or broken items not as trash, but as things that can be repaired, reused, recycled, composted, or disposed of to create energy—turning them from problems into part of the solution.

To transport radioactive waste deep underground, rail systems must first be constructed in proposed sites such as Yucca Mountain.

Wasteful Behavior

Founded in 1953, Keep America Beautiful is dedicated to reducing waste, preventing litter, and beautifying the American landscape through educational activities and hands-on initiatives. Every year, the organization sponsors the Great American Cleanup. From March 1 to May 31, approximately three million volunteers donate their time to cleaning beaches, collecting litter, holding recycling drives, and planting trees and flowers in their communities. In 2008, the Great American Cleanup collected 86 million pounds (39 million kg) of litter from roadsides, shorelines, public parks, and other areas. To find out how you can get involved in the next Great American Cleanup, visit www.kab.org.

The beauty of nature is too often obstructed by waste, but individuals and organizations can change that by picking up the trash.

Glossary

aquifers—underground bodies of rock, sand, or gravel that can hold or transmit water

atmosphere—the layer of gases that surrounds Earth

corrosive—tending to cause corrosion, or the gradual wearing away of another substance by chemical action; acids are one type of corrosive substance

developing countries—the poorest countries of the world, which are generally characterized by a lack of health care, nutrition, education, and industry; most developing countries are in Africa, Asia, and Latin America

endocrine system—a body system that is made up of endocrine glands, such as the thyroid and pancreas, that send hormones into the bloodstream

geotextiles—landscaping fabrics made out of polyester or other materials that improve drainage and soil stability

humus—the brown or black organic component of soil, which consists of decayed or decomposed plant and animal substances; humus adds nutrients to soil and helps it retain water

Industrial Revolution—a period during the late 18th and early 19th centuries in Europe and the U.S., marked by a shift from economies based on agriculture and handicraft to ones dominated by mechanized production in factories

infertility—the state of being unable to have children

irrigation—the distribution of water to land or crops to help plant growth

liquefied natural gas—a form of natural gas that has been subjected to pressure and then cooled in order to form a liquid; it can be used for heating or cooking or to fuel vehicles

nomadic—characteristic of people who do not settle in one location but move from place to place, usually seasonally

nuclear—having to do with energy created when atoms are split or joined together

organic—derived from or relating to living matter

pathogens—disease-causing agents, such as bacteria or viruses

peddlers—people who travel from place to place, selling items

pesticides—substances used to kill insects or other living things that are harmful to certain plants or animals

plasma—a superheated gas that is a good conductor of electricity

radioactive—characteristic of substances such as uranium that give off particles of energy as their atoms decay; the energy is dangerous to human health

reactive—having a tendency to react chemically, taking part in a chemical reaction with another substance; when hazardous wastes react, they may cause an explosion or create a poisonous gas

scrubbers—devices that remove pollutants from a gas

urban—related to or living in a city

volatile organic compounds—chemical compounds present in substances such as paint that contain the element carbon and that evaporate at low temperatures, releasing gases that cause air pollution

Bibliography

Ackerman, Frank. *Why Do We Recycle? Markets, Values, and Public Policy.*
Washington, D.C.: Island Press, 1997.

Grossman, Elizabeth. *High Tech Trash: Digital Devices, Hidden Toxics,
and Human Health.* Washington, D.C.: Island Press, 2006.

Kostigen, Thomas. *You Are Here: Exposing the Vital Link Between What We
Do and What That Does to Our Planet.* New York: HarperOne, 2008.

McCorquodale, Duncan, and Cigalle Hanaor, eds. *Recycle: The
Essential Guide.* London: Black Dog Publishing, 2006.

Popov, V., A. G. Kungolos, C. A. Brebbia, H. Itoh, eds. *Waste Management
and the Environment III.* Boston: WIT Press, 2006.

Royte, Elizabeth. *Garbage Land: On the Secret Trail of Trash.*
New York: Little, Brown and Company, 2005.

Strasser, Susan. *Waste and Want: A Social History of Trash.*
New York: Metropolitan Books, 1999.

Vandenbosch, Robert, and Susanne E. Vandenbosch. *Nuclear Waste Stalemate: Political
and Scientific Controversies.* Salt Lake City: University of Utah Press, 2007.

For Further Information

Books

Bowden, Rob. *Waste, Recycling, and Reuse: Our Impact on the Planet.*
Austin, Tex.: Raintree Steck-Vaughn Publishers, 2002.

Dorion, Christiane. *Earth's Garbage Crisis.*
Milwaukee: World Almanac Library, 2007.

Foran, Jill, ed. *A Planet Choking on Waste.*
North Mankato, Minn.: Weigl Educational Publishers, 2003.

Gifford, Clive. *Waste.* Chicago: Heinemann Library, 2006.

Web Sites

Keep America Beautiful Presents: Clean Sweep, USA
http://www.cleansweepusa.org

Environmental Protection Agency: Recycle City
http://www.epa.gov/recyclecity/mainmap.htm

Waste Watch: Recyclezone
http://www.recyclezone.org.uk

Washington State Department of Ecology: Kids Recycle Page
http://www.ecy.wa.gov/programs/swfa/kidsPage

Index